LIGHT & COLOR
Images from New Mexico

Masterpieces from the Collection of
the Museum of Fine Arts, Museum of New Mexico

Museum of New Mexico Press
Santa Fe

ACKNOWLEDGEMENTS

The creation of the touring exhibit, "Light and Color: Images from New Mexico," is the result of the outstanding cooperation of two important American art institutions: the Museum of Fine Arts in Santa Fe, New Mexico, and the Sheldon Art Gallery of the University of Nebraska at Lincoln. While the Museum of Fine Arts is undergoing renovation, Norman Geske, Director of the Sheldon Art Gallery, conceived the idea of continuing to make part of its superb collection of paintings available to the public through a touring exhibit to the Mid-America region. The Museum of Fine Arts generously allowed him great freedom in selecting the paintings from its holdings which comprise the exhibit. Mid-America Arts Alliance is delighted to be the vehicle for this model cooperative effort.

We are immensely grateful for Norman Geske's initiative on behalf of the Mid-America region which has made this exhibit possible. We also recognize the expert knowledge and painstaking care with which he and Jon Nelson, Assistant Director of Sheldon Art Gallery, selected the paintings for "Light and Color." Our thanks, too, go to the staff of the Sheldon Gallery, who have cared for and stored these paintings prior to the tour.

In his splendid introduction to this catalogue, Mr. Geske has thanked specific staff members of the Museum of Fine Arts for their help. On behalf of Mid-America Arts Alliance, I would like to add our thanks to Ellen Bradbury, Director of the Museum of Fine Arts in Santa Fe, and her staff for their fine cooperation.

We are grateful to the National Endowment for the Arts' Museum Division for its exhibition grant to M-AAA to assist this tour.

David Smith, Jerry Kramer, and Gary Goldberg of the M-AAA Visual Arts staff deserve recognition for their hard work in handling all the logistics of the exhibit's tour.

Finally for their support of this exhibit, we thank the Mid-America Arts Alliance Board of Directors and our region's five state arts agencies: the Arkansas Arts Council, Kansas Arts Commission, Missouri Arts Council, Nebraska Arts Council, and the State Arts Council of Oklahoma.

Henry Moran
Executive Director
Mid-America Arts Alliance

FOREWORD

...the sun sank lower, a sweep of red carnelian-colored hills lying at the foot of the mountains came into view; they curved like two arms about a depression in the plain; and in that depression was Santa Fe. A thin, wavering adobe town...a green plaza...at one end a church with two earthen towers that rose high above the flatness. The long main street began at the church, the town seemed to flow from it like a stream from a spring. The church towers, and all the low adobe houses, were rose colour in that light...

So Willa Cather, in **Death Comes for the Archbishop,** describes Bishop Lamy's first visit to New Mexico's capital city.

New Mexico has been discovered and rediscovered through the centuries. The canyons and valleys with their small streams attracted the Paleo-Indians who camped there and carved pictographs into shady cliff walls. In 1540, Coronado led his Spanish soldiers into the Indian pueblo of Zuni and claimed New Mexico for Charles V of Spain. Settlers followed, and in the winter of 1609 a group led by Don Pedro de Peralta settled near the present site of the Fine Arts Museum in Santa Fe. Small Spanish communities spread out from Santa Fe and along the trails that connected them with the United States and Mexico.

Around the turn of the twentieth century the first artists came to New Mexico seeking exotic subjects to paint. Indians and Spanish people provided a foreign atmosphere which was as satisfying to late nineteenth-century taste as the popular subjects of Bavarian peasants or Algerian markets. The pleasant climate and rough but inexpensive life convinced some artists to stay, particularly in Taos and Santa Fe. They built houses of adobe, grew fond of chili and developed a scorn for Eastern life and comforts. More importantly, although these first artists had Eastern and even European training, they reacted to the light and color and the simplified land forms of the Southwest.

In 1907 a small art gallery was created in the old Palace of the Governors on the Santa Fe plaza by Dr. Edgar L. Hewett, the first Director of the Museum of New Mexico. As he wanted to encourage and augment the small community of local artists, Hewett also provided some of them with free studio space in the back of the ancient Palace. When larger and more permanent quarters were required, the territorial government agreed to fund an art museum; in 1910 Hewett embarked on a building program. In 1917 the structure, a synthesis of the old mission church at Acoma with general elements of adobe architecture, opened with great fanfare. The building has become a prototype for what is called Santa Fe or Pueblo style architecture.

Hewett had persuaded artist Robert Henri to visit New Mexico in 1916, and Henri's warm personality and democratic philosophy pervaded the new museum. Henri advocated an "open door" policy — any artist who wanted to do so could come in and hang his or her work on the Museum's walls. In its early years the Fine Arts Museum acted as a discussion center where artists met and were welcome. It was a radical and an endearing beginning for an institution.

In 1977 the Museum of Fine Arts received an appropriation from the State of New Mexico for renovation and expansion. The permanent collection presently contains over 10,000 paintings, prints, photographs, and sculptures which have been donated and purchased over the years. Rather than put the entire collection into storage while the Museum was temporarily closed, Norman Geske, Director of the Sheldon Memorial Art Gallery, and I decided that selected works should go on tour. With this in mind the collection was reviewed by the Museum of Fine Arts staff and Mr. Geske. The high quality of the works impressed us all. We are delighted to be able to share these masterpieces of the collection — reflections of the light and color of New Mexico.

Ellen Bradbury, Director, Museum of Fine Arts, Santa Fe, New Mexico

INTRODUCTION

This exhibition came into being as the result of very special circumstances. In March of 1980 the Museum of Fine Arts in Santa Fe, one of the component parts of the Museum of New Mexico, closed its doors for the first time in its sixty-three year history in order to accommodate the expansion and renovation of its facilities. This situation produced an immediate need for a place to which the museum's extensive collections might be removed for the duration of the reconstruction. A storage warehouse might have been the simplest and most obvious solution but it would have removed from view a considerable number of important works of art, a body of work which, in fact, constitutes one of the best collections of its kind, the visual archive of a region of our country which, perhaps more than any other, has contributed in an important way to our national art.

Fortuitous circumstances found the Director and the Assistant Director of the Sheldon Gallery in Santa Fe at the very moment when this problem had reached its moment of decision. The possibility of putting a selection of the Museum's pictures on the road, on a kind of good will tour, for the period of the renovation suggested itself as an opportunity not to be missed, providing the means of making such a tour work in practical terms could be found. The means were found, fortunately, in the visual arts program of the Mid-America Arts Alliance and what, at the outset, had been hardly more than the inspiration of a moment, moved toward the reality of this exhibition.

It is probably unprecedented that the director of one museum has been turned loose in the storerooms of a colleague's museum to choose fifty works of the greatest personal interest. The only restraints had to do with the previous commitment of certain examples to other institutions and the ever present matter of condition.

What a feast of self indulgence, to select what one likes best, without reference to any preconceived notion of what the exhibition should be.

Not too surprisingly, the result has in it a substantial piece of the history of painting in New Mexico. It does include works by many of the best known artists who have worked there, but it is probably better that the exhibition be looked at in a somewhat different context. It is, strictly speaking, an exhibition of the very idea of an art museum, the bringing together by purchase, gift, bequest, loan or accident of works of art considered to be of genuine quality by their creators or owners, or by the trustees and professional staff of the institution. Like most museum collections it is not "complete" in its representation of its chosen area of concern and it does not consist of masterpieces alone. Some of the pictures are "early" or atypical or incidental. Some do not even depict New Mexico, having been done before or after the artist's experience of the region. In a broader sense, however, taking the whole as representative of American painting in the twentieth century, it provides an illuminating review of our national art history as it developed within the confines of a very special geographical region. Everything that characterizes modern art is represented. Realism and Surrealism, Impressionism and Expressionism, Cubism and Constructivism, the documentary, the poetic and the mystical.

The one descriptive term conspicuous by its absence in the above is **regional**, and although what has already been said about these pictures may indeed explain its absence it is still to be dealt with. Certainly regionalism as we know it in the history of art is a limiting term in its application to subject matter, to style and to point of view. Nothing in this exhibition precludes its use, yet it still must be acknowledged that among all the special places in our country which have become famous as artists' colonies, Provincetown, Woodstock, Carmel and others, not one of them has provided a strictly comparable mixture of ethnic, social and environmental elements. Three distinctively different cultures, the Indian, the Spanish and the "Anglo," have come into an historic conjunction with a result that can only be described as unusual.

Perhaps the best word to describe the quality of this place which has exerted so persuasive an influence on the artists who have lived there is **ambience,** which somehow seems to mean something more than environment and something different than atmosphere. Its meaning is in part geographical, part cultural, part psychological. Three cultures, each one different and enormously complex in itself, exist in an almost miraculous state of balance in New Mexico. In this place the artist, whether Indian, Spanish or Anglo, seems to work with a greater sense of freedom and ease.

The Museum of Fine Arts, which is only one of several divisions of the Museum of New Mexico, is perhaps unique in one respect at least among American art museums. It is the only one which came into being in direct response to the presence and activity of artists drawn to the region by the word of mouth news that here was a wonderful place to work. The word was that here was a locale alive with light and color in a varied and dramatic landscape where a remarkable harmony existed between the inhabitants and the environment. The museum was established to provide a place where the creativity inspired by this remarkable ambience might be displayed for the further enjoyment and inspiration of the increasing numbers of people drawn to this extraordinary country.

It appears that there was no immediate sense of obligation on the part of the museum's founders to record the artistic history of the region, although it was certainly not long before that task became an important part of the museum's function. Similarly, and here is where Santa Fe differs significantly from most other communities, there was no effort to import and impose the artistic tradition of the Old World. Founded in 1907, its adobe style building was completed and dedicated in 1917, one of the first art museums in the United States devoted exclusively to the work of contemporary artists.

In the years since, the Taos/Santa Fe school has acquired historical status and now provides only the starting point for the museum's collections. Also, in the years since, the creative community in New Mexico has spread far beyond the municipal limits of Taos and Santa Fe, but the museum has maintained its basic interest in today's art. The growing obligations of museum keeping, as the collections have expanded to accommodate sculpture, graphics, photography and the burgeoning activity of native American artists, have not put a greater distance between the museum and the creative community. The Museum of Fine Arts has been constantly aware of the ever enlarging scope of artistic activity in the Southwest, which, as has already been noted, is less regional than ever before and has become, in fact, a reliable gauge of the national scene.

The comments which follow are based on personal thoughts and reactions; hopefully they provide a kind of conversational gallery tour which makes no attempt to tell the whole story or even to mention every work included. At the end of the catalog the brief bibliography which exists for the history of modern art in New Mexico is provided for those who desire more information than can be provided here.

The earliest painting in the exhibition is certainly "Old Mesilla Plaza" (page 21), lent to the Museum by the Smithsonian Institution. The precise date of execution, probably about 1865, is unknown. For us the picture's importance and its appeal is in its almost complete lack of artistic style or imaginative glamour. It says to us quite clearly that a large part of such considerations are, after all, in the eye of the beholder and what we have here is the "Land of Enchantment" seen with the cold and anxious eye of the pioneer and described in a plain and somewhat fumbling recital of facts. It is interesting to note that neither picturesque natives nor dramatic landscape seem to be of importance to this painter who is only concerned with recording the recognizable proofs of a civilized presence. Even so there is always something about the first human record of a given place, however meager or inadequate, which is peculiarly exciting. Our own seeing is somehow preempted by that first perception, and our own subsequent knowledge of the subject is forever conditioned by it.

The Taos Society of Artists, founded with six members in 1912 and enlarged to ten with a final addition in 1927, is a story unto itself. (See note at end of bibliography.) The group included Joseph H. Sharp, Eanger I. Couse, Ernest L. Blumenschein, Oscar E. Berninghaus, Bert G. Phillips, Herbert Dunton, Walter Ufer, Victor Higgins, Catherine Critcher and Kenneth Adams. It should be noted that the Society was organized primarily for promotional purposes outside New Mexico, where all the members won prestigious prizes in the nation's major exhibitions. Several were patronized by the Santa Fe Railroad, which used their paintings to encourage tourism. When the State of Missouri sought artists to decorate the new capitol building in Jefferson City, Blumenschein, Berninghaus and Dunton among others were chosen to do the work. Few groups of painters in our history have achieved so substantial a degree of success both with the general public and with private collectors. Few public collections, with the possible exception of that of the Gilcrease Institute of Tulsa, have achieved the completeness in representation of the Taos Society of Artists to be found in private collections such as the Anschutz and Harmsen Collections in Denver, the Adkins Collection in Tulsa, the Eiteljorg Collection in Indianapolis and the Stark Collection in Orange, Texas. Admittedly, long before the death of the last of the Society's members, Adams in 1966, their prestige had come to be limited to that part of the art public which has been resistant to the blandishments of modern art. They have, in a sense, dropped out of our art history to the accompaniment of words of critical derogation such as **academic, illustrative** and **provincial.**

To digress for a moment, one of the worst snobberies of modern art, rampant in recent decades, has been the notion of the mainstream which, if it can be identified at all, assumes a single line of development moving toward a theoretical ultimate. When used as a standard this notion excludes everything that doesn't fit. In this situation, in the middle decades of the century, the art and artists of Taos and Santa Fe did not appear to fit. It now seems possible that a larger view of the development of the art of our time might find that these painters are valid members of the same artistic society that produced all the other variants, digressions and experiments that characterize twentieth century art.

In looking at the work of the Taos Society of Artists it is entirely proper and necessary to recognize their common source in the academic tradition of the nineteenth century, in the art schools of Chicago, Cincinnati, Paris and Munich, and to acknowledge the range and popularity of their activity as illustrators and commercial artists. This background accounts for their skill in drawing and composition and for their ability to summarize a subject in the most effective way. They are all romantics at heart, with occasional digressions into a vigorous realism tempered by the light and color of the region. It is also possible to see Impressionism in their work, as in Ufer's "Wild Plum Blossoms" (page 86) and Blumenschein's "Adobe Church" (page 35), but, admitting that these artists could hardly have been unaware of Impressionism in the hands of its masters, it seems more likely that it served them only as a useful precedent for their own individual development.

Ernest Blumenschein was an exact and sympathetic observer of Indian life, interested in the creation of an image which, in all its parts, color, drawing and composition, would convey an emotional as well as an objective reality (page 34). There is usually a pervasive sense of drama in his work and frequently an inventive manipulation of space which gives his imagery a high degree of purely visual excitement. The charming "Adobe Church" (page 35) is an unusual example of the artist's work in the simplicity of the subject and the spacious ease of its execution, but it shows, even so, his habitual concern for a tightly composed pattern of forms, precisely seen, but delineated with the fluency and freedom of a sketch.

Oscar Berninghaus was not only one of the most successful of the Taos painters but was one of the most skillful as well. Paintings from the early part of his career are noted for their panoramic sweep and fine detail. "Corn Dance Day" (page 31) is one of these works, full of the sensory quality of the event, crowded, dusty and noisy. Notable too is his sense of the bleaching light which envelops and unifies the activity of the scene. Berninghaus was not as well known for his portraits as other Taos painters and in this instance it can be noted that the extraordinary detail in the drawing of the picture does not include a single face in the crowd. Later paintings such as "Portrait of A Rabbit Hunter" (page 30) show him paying greater attention to large, single forms in the foreground of his subject, developed again with incisive detail and a marked textural enrichment in the paint surface. His skill as a mural painter is evidenced in the way foreground and background elements are organized in a counterpoint of movement across the picture plane.

Walter Ufer is perhaps the prototypical member of the Taos/Santa Fe school. Quite apart from his artistic skill he was enormously gifted in the social graces of the profession, and was, unfortunately, the victim of his own abilities. Van Deren Coke has summarized Ufer's quality as a painter very well: "Never taking an intellectual approach, he assembled the components of his pictures in such a manner that little was left to the imagination. There was no symbolism or evidence of introspection, nothing of the objectivity of Berninghaus or Sharp, but rather an art that depended upon the subject to speak for itself. Ufer seemed fascinated by his ability to paint easily recognizable forms in an anecdotal manner."[1] Even more unfortunately he was persuaded by the demands of his market to settle for one picture, the definitive Ufer, which contains Indians, white horses and a mountain setting. "Chance Encounter" (page 87) would appear to be one of these trademark pictures, which, in the end, destroyed his reputation and ultimately the man himself. It is gratifying, however, to see an atypical work such as "Wild Plum Blossoms" (page 86), which indicates that he was capable of a simple, painterly response to simple subjects.

Bert Phillips' "Our Washerwoman's Family" (page 75) is a masterpiece of domestic genre painting, completely within the traditions of the nineteenth century academy. A strong vein of sentiment and a measure of idealization make his work as a whole a less than factual record of the people for whom he felt so strong an attachment. For the contemporary viewer, for whom an open display of sentiment has no attraction, there remains a remarkable sense of color, refined and luminous.

In the "Green Corn Dance" by William Penhallow Henderson (page 51), we have an example of the shift away from the primarily documentary impulse toward a more personally expressive style. Henderson and others who based themselves in Santa Fe were affected directly, or indirectly, by the "modernism" which had entered the nation's artistic bloodstream after the Armory Show in 1913. In his paintings (he was also active as an architect and sculptor) Henderson presented an essentially emotional record of standard subject matter; the Indian dances and the Penitente processions. Rich color and a highly rhythmic organization of forms are the vehicle for a distinctly personal interpretation of these subjects.

Of all the members of the Taos Society of Artists, Victor Higgins was the only one who showed a capacity for growth which eventually led him away from the group's "official" subject matter into a more personal exercise of his abilities.

In this exhibition there is the opportunity to see Victor Higgins in three different roles, as the composer of large scale figure subjects, the portrait painter and the landscape painter. In each the distinct personality of the man is evident in the bright luminosity of his color and a notably sensuous rapport with his media, oil and watercolor.

In the "Winged Victory" (page 57), an early pre-New Mexico example, there is an inventive sensibility at work, testing the validity of the rules and speculating on the possible alternatives to the academic formulae which underlie the image.

Similarly, the portrait of "Mercedes First Communion" (page 56) is, at heart, a Santo, simple, direct and symbolic in its unassuming sweetness, framed as a Santo might be in an aureole of petal forms very much like the tin leaves that frame the Santo visible in Bert Phillip's "La Lavanderia."

In "Landscape, Taos, New Mexico" (page 57), we have the artist working in a manner very like that of John Marin. It would be simple to ascribe the quality of this painting to the influence of Marin, were it not for the persistent presence of Higgins in the bright, lyric tone of the picture which is quite unlike the crisp brusqueness of Marin's manner.

One of the historical considerations concerning the role of Taos/Santa Fe in American art, which has until recently gone without investigation, is its relationship to the development of "The Eight" in Philadelphia and New York. It could be argued with considerable rightness that the careers of the individual members of the Taos Society of Artists were in large part motivated by the same impulse toward an expressive realism which is so much a part of the point of view of the Ash Can painters. This similarity is indicated in the enthusiasm with which the New Yorkers, Henri, Sloan, Davey, Bellows and Kroll, brought to their experience of New Mexico. They were not particularly interested in the documentation of Indian culture or the Old West, but, aside from such concerns, their response was comparable in artistic terms to that of their predecessors and contemporaries in New Mexico. The important link between them is their common involvement with the sensory properties of the region, the light and color which are the germinal elements of these images. A symbolic link is provided in the fact of Henri's election to an "Associate" membership in the Taos Society of Artists in 1918.

New Mexico provided a considerable stimulus to the work of Robert Henri. There, in contrast to the urban tonalities of New York, he found a full range of the brilliant color which was his principal concern throughout his career. His devotion to color as the vehicle of form and his empathic response to the personalities of his favorite subjects, children, dancers, gypsies, picturesque eccentrics and "exotics," found complete expression. "Dieguito" (page 55) is one of the unqualified master works in this collection and certainly one of the artist's finest works. It is a superbly bravura performance of the painter's art. If it lacks the ethnographic value that characterizes the portraits of Joseph Sharp, it achieves something else in the expression of a personality, vibrant and alive, before your very eyes.

In the summer of 1917 Leon Kroll and George Bellows spent some weeks with Robert Henri, who had urged them to come to Santa Fe to see for themselves the wonders of the place. Bellows' reaction was recorded in two canvases which stand out in his total work as exceptional for their warmth and fluency. Here, perhaps more than is usually the case in his work, he has immersed himself in sensory experience. The sky in

"Chimayo" (page 27) is a tour de force in its depiction of the light peculiar to an imminent thunderstorm. The animals, the human figures, are immobile presences, transfixed by the tension of the moment. It is of some interest to note that in this scene, all light and atmosphere, the one note of positive color which focuses the picture is the vegetable red of the chilis hung in the porch of the house. It is also worth noting that one of Bellows' best known works, "The White Horse" of 1922 in the collection of the Worcester Museum, contains the same dramatic sky, to say nothing of the same horse, and might perhaps be seen as a kind of recollection of New Mexico.

Kroll's record of his visit is possibly contained in this single work, "Santa Fe Hills" (page 65). The landscape forms, buildings, human, animals and a turbulent sky are handled in accordance with the teaching of Robert Henri. His palette is not strictly speaking of the place, but is his own, with the characteristic range of blues, grays, greens and violets which is to be found in his work throughout his career.

As one of the principal disciples of Robert Henri, Randall Davey displays a typical devotion to the **alla prima** procedure which prefers the spontaneous drawing of forms and effects without preliminaries, subsequent modeling or correction. This approach accounts, in large part, for the freshness of the pigment in his pictures (pages 42 & 43). Our enjoyment of his enjoyment should not, however, blind us to the fact that it requires an unusual degree of discipline, both optical and manual, on the part of the painter; an ability to see, understand and place the forms in their proper relationship on hardly more than the intuition of the moment.

Of all the painters from the East few ever managed so complete an identification with New Mexico as John Sloan. Drawn there first of all by the urging of Robert Henri, he became not only a regular summer resident but also a figure of influence on the local scene. The Museum of New Mexico's initial policy of an open door to all artists who wished to exhibit was, at least in part, Sloan's doing. It was the very policy he sought to establish in New York, and his was certainly the most impassioned outcry when that policy was abandoned in Santa Fe. In any case, Sloan's work in New Mexico includes a considerable number of his finest works, among them "Ancestral Spirits" (page 83).

In this painting all the artist's interest in spontaneous action is given full rein. This is a very different view of Indian life than was presented in earlier works of the Taos/Santa Fe artists. First of all it is not "distanced." There is no sense of the artist being a detached observer. Sloan seems to be as close as he can get, almost involved in the procession itself. It is wonderfully immediate.

In the Twenties the pattern of artistic activity in New Mexico became increasingly diverse. There was a new organization on the scene, a group of painters calling themselves **Los Cinco Pintores.** This group included Josef Bakos, Fremont Ellis, Walter Mruk, Willard Nash and Will Shuster. Their declared purpose was "to take art to the people and not to surrender to commercialism," which as a statement of purpose suggests something of a reaction to the commercial prestige of the Taos Society of Artists. Their work as a group was diverse to a degree, but it was plain that their central interest was no longer the documentation of a vanishing culture, but was, instead, focused on the making of pictures expressive of their own personal points of view with less emphasis on the specifics of subject matter.

Of the works of these men, Bakos and Ellis and Nash, included in this exhibition, one can only say that their "modernism" which came as something of a shock to the New Mexico scene now seems conservative enough (pages 23, 44, 45 & 69). As was the case with many

American artists of their generation, the lessons of the Armory show were adapted rather than assimilated. Cezanne, Impressionism, Cubism and Post-Impressionism were all given an American form wherein each was held back from the full thrust of their stylistic implications.

B.J.O. Nordfeldt's work in New Mexico has the heightened color and boldness in brushwork which one might expect, but he was also something of a cosmopolite, taking his inspirations from an international range of sources.

His "Antelope Dance" (page 71) is the most ambitious of his several uses of Indian dance and ritual as a theme, but it also takes on the character of a challenge to its obvious prototypes in the work of Cezanne. One is reminded that Hartley also indulged in a similar challenge in his landscapes painted in Provence in 1925-1927. Comparison is, of course, pointless. Cezanne is not Nordfeldt, nor Nordfeldt Cezanne, but to undertake the use of a famous compositional prototype, a purely formal structure of landscape, trees and mountain as the setting for an emotionally charged dance ritual, with the symbolic presence of the Black Mesa substituted for Mt. Ste Victoire, is, at the very least, impressively audacious.

Gustave Baumann is of special interest in that he, like the members of the Taos Society of Artists before him, came out of a background of European academicism and the conventional regionalism of Indiana's Brown County School of painting, and, in his case, New Mexico brought about important changes in the character of his work. He is unquestionably one of the masters of the color woodcut in twentieth century art, and his prints, which were of sufficient fame to draw admiring students and collectors from Japan, are certainly among the definitive images of New Mexico. Equally remarkable although less well known are his paintings, which probably are the first by an outsider to draw upon the formal precedents of Indian art.

In such works as "Winter Ceremony, Deer Dance" (page 25), the idea of space is contained in an abstract field of color, richly textured, imprecise, intangible, much like the spaces in Chinese painting which are described by mist or cloud. There is also the reflection of Baumann's study of the Indian petroglyphs in Frijoles Canyon. His essential distinction was in his ability to see and understand his subjects in terms other than those of traditional academic art, in terms which are drawn from the visual traditions of the region itself.

Of all the painters who came to New Mexico at the urging of Mabel Dodge Luhan, including Maurice Sterne, Andrew Dasburg and John Marin, Marsden Hartley was perhaps the least at home there. In his case we have a strongly individualized personality deeply affected by the experiments of modern painting in Europe before he arrived in New Mexico. In the Southwest, however, it appears that the bold, assertive character of the environment was in competition with his own highly idiosyncratic patterns of thinking and working. It seems that he was not really at ease in the time he spent there. His aristocratic distaste for the art world of Taos is recorded. Even so, a considerable number of landscapes and still life compositions reflect the depth of his response: "any one of these beautiful arroyos and canyons is a living example of the splendour of the ages...and I am bewitched with their magnificence and their austerity; as for the color, it is of course the only place in America where true color exists, excepting the short autumnal season in New England."[2] It is significant perhaps that Hartley's most important use of New Mexico took the form of a series of works entitled "Recollections of New Mexico," painted in Europe three years after his departure. In the museum's superb "El Santo," which, unfortunately, could not be included in this exhibition, his rapport with the emotional mood of his subject is clearly evident. In the simpler "Still Life" (page 49), we have a painting which, in all probability, was not executed in New Mexico, but it still reflects the sombre earth tones and simplification of forms which were part of his response to the place.

The death of Andrew Dasburg in 1979 marked the end of a period which witnessed the assimilation of all the major innovations of twentieth century art. A participant in the development of Cubism at the beginning of the century, he completed a long life of continuous activity on his own part and of constant encouragement and support of younger artists who came to know him. His own growth, in refinement of vision and style, was a guiding phenomenon for several decades. At the moment of this exhibition an important retrospective of Dasburg's work is being circulated by the University Art Museum at Albuquerque, and, naturally most of his available work is included. We can, however, draw attention to his skill as a portrait painter in the imposing image of "Charles Augustus Ficke" (page 41). Aside from the likeness, the picture is a formal design of extraordinary quality, with volume and surface and edge orchestrated with consummate skill in a definitive example of American Cubism.

It is perhaps a little unfair to represent Tom Benrimo with a work which comes from the initial place of his activity in New Mexico, except for the fact that his deeply personal sense of fantasy and visual poetics is clearly present in this painting which is in the manner of Daliesque Surrealism (page 29). It should be noted that in his later work Benrimo developed a very different style in which his subjectivity was expressed in a more personal way.

It is difficult to consider Dorothy Brett the painter against the competitive knowledge of her association with D. H. Lawrence. Already a painter on her way at the time of their first meeting, it appears that her personality as a painter was to some extent frustrated until after her return to permanent residence in Taos after Lawrence's death in 1930. There she began the remarkable series of paintings of Indian dances and ceremonials which constitute her best claim to our attention. Of her interest in Indian subjects she has stated: "It is difficult to know from what one paints, whence it comes, or why it comes out in the particular technique that it does. Perhaps there is no way to explain why I, for instance, have chosen to paint Indians, rather than landscapes or flowers or abstractions, except that in my childhood I fell in love with one of Buffalo Bill's Indians. He rode wildly around the arena — naked; painted lemon yellow; a great war-bonnet with its feathers cascading down from his head to his horse's feet; screaming and yelling; firing off his gun. This heroic figure I have never forgotten, though I was only five years old."[3]

Any interpretation of Dorothy Brett's "The First Born" (page 37) is hazardous without a more extended exploration of meaning than is possible here. What is clear is the intensity of the appeal to our perceptions. Brett was not a conventional artist — at times an only barely competent artist, but her simplicity is at the same time her strength and her vision of New Mexico is unique.

The exhibition includes the work of twelve living painters, all, with one exception, still working there. This group of paintings provides ample evidence that the area continues today, as it has for almost a hundred years, to draw talented individuals into its spell. Today, of course, the historical dimension of the "Old West" is dormant if not dead in the practice of the professionals who live there, and the presence of the new academic art of the universities and a multimillion dollar art market have made a very different world. We are far from the simple documentary impulse of Joseph Sharp. Instead, we have an artistic practice that embraces physical science, psychology and metaphysics.

Of the contemporaries in the exhibition it can be noted that the observed realities of New Mexico are certainly discernible in the work of Peter Hurd (page 59), Earl Stroh (page 85), Elmer Schooley (page 79) and John Wenger (page 93), but, it is a reality seen through the subjective lens of the artist's personality.

In the paintings of Raymond Jonson (page 63), Lee Mullican (page 67), Clinton Adams (page 17) and Marcia Oliver (page 73), the sensibility of the artist has translated the experience of phenomena into the language of abstraction, and in the work of Douglas Johnson (page 61) we find an unusual fusion of realism and abstraction.

Of special interest are the paintings of Fritz Scholder (page 77) and R. C. Gorman (page 47), in that they represent the emergence, in the native art of the region, of a contemporary point of view. The erasure, by talents like these, of the line of distinction between their art and ours may well be the most important development of all.

First of all may I thank the staff of the Museum of Fine Arts in Santa Fe, Ellen Bradbury, Don Humphrey, Steve Yates and Sandra Demilio, for their help and endless patience. And finally, I would like to acknowledge again my colleague, Jon Nelson, who participated in a substantial way in the selection of the paintings for the exhibition, and David Smith, Jerry Kramer and Gary Goldberg, of the Visual Arts Program of the Mid-America Arts Alliance, all of whom contributed to the logistical accomplishment of the exhibition and its catalog. The Sheldon Gallery is much indebted to the Alliance for the opportunity to show these pictures in Lincoln prior to their tour under Mid-America's auspices. For the record, "Light and Color: Images from New Mexico" was shown at the Sheldon Memorial Art Gallery, University of Nebraska-Lincoln, from June 17 through August 3, 1980.

Norman A. Geske, Director
Sheldon Memorial Art Gallery
University of Nebraska-Lincoln

Footnotes:
1. Van Deren Coke, **Taos and Santa Fe, The Artist's Environment 1882-1942,** Albuquerque, University of New Mexico Press, 1963, p. 24.
2. Hartley to Harriet Monroe, September 13, 1918, **Archives of American Art.**
3. Dorothy Brett, "Painting Indians," **New Mexico Quarterly,** Vol. XXI, Summer, 1951, No. 2, pp. 167-168.

Selected Bibliography:
Bickerstaff, Laura M., **Pioneer Artists of Taos,** Denver, Sage Books, 1955 (out of print).
Coke, Van Deren, **Taos and Santa Fe — The Artist's Environment 1882-1942,** Albuquerque, University of New Mexico Press, 1963 (out of print).
Robertson, Edna and Nestor, Sarah, **Artists of the Canyons and Caminos,** Layton, Utah, Peregrine Smith, 1976.
Trenton, Pat, **Picturesque Images from Taos and Santa Fe,** Denver Art Museum, 1974.

Note: A history of the Taos Society of Artists by Patricia J. Broder will be published in the fall of 1980 under the title, **Taos: A Painter's Dream.** It will be a New York Graphic Society publication from Little, Brown & Co., Boston.

CLINTON ADAMS

EVENING SEQUENCE,
1968,
acrylic on canvas
60-1/4" x 36-5/16"
(152.8cm. x 92.2cm.)
Gift of the artist

Born in Glendale, California, 1918. Studied at University of California, Los Angeles. Has served on the faculties of the Universities of California at Los Angeles, Kentucky and Florida. Presently Dean of the College of Fine Arts at the University of New Mexico. Lives in Albuquerque, New Mexico.

Solo Exhibitions: Roswell Museum and Art Center, 1971; University of New Mexico Art Museum, Albuquerque, 1972.

Public Collections: Achenbach Foundation for Graphic Arts, San Francisco; Amon Carter Museum of Western Art, Ft. Worth; Art Institute of Chicago; Grunwald Graphic Arts Center, University of California, Los Angeles; Museum of Modern Art, New York; Sheldon Memorial Art Gallery, University of Nebraska-Lincoln.

KENNETH ADAMS

HOUSE IN THE SUN,
watercolor on paper
(sight) 9-3/4″ x 17-3/8″
(24.8cm. x 43.6cm.)
Gift of Judge and Mrs.
Oliver Seth, 1977

Born in Topeka, Kansas, 1897.
Studied at Art Institute of
Chicago; Art Students League,
New York with Kenneth Hayes
Miller, Maurice Sterne, Eugene
Speicher and Andrew Dasburg.
Taught at the University of New
Mexico, 1933; 1938-63. Elected
an Associate of the National
Academy of Design in 1938 and
Academician in 1961. In 1927 he
was the youngest and last
member elected to the Taos
Society of Artists. Died in
Albuquerque, 1966.

Solo Exhibition: University of
New Mexico Art Museum,
Albuquerque, 1964.

Public Collections: Eiteljorg
Collection, Indianapolis;
Harwood Foundation of the
University of New Mexico, Taos;
Joslyn Art Museum, Omaha;
Kansas State University,
Manhattan; United States
Department of Labor,
Washington, D.C.; University of
New Mexico, Albuquerque.

Kenneth M. Adams –

OLD MESILLA PLAZA,
oil
29-7/8″ x 48-5/8″
(75.9cm. x 123cm.)
Courtesy of the National
Collection of Fine Arts,
Smithsonian Institution,
Washington, D.C.

JOSEF BAKOS

GERANIUMS,
oil
40″ x 30-3/8″
(101.5cm. x 76.6cm.)
Gift of Friends of the
Museum

Born in Buffalo, New York,
1891. Studied at Albright Art
School, Buffalo and in Denver
with John E. Thompson. Taught
at the University of Colorado,
University of Denver and Santa
Fe High School. Arrived in
Santa Fe, 1921, one of the
original **Los Cinco Pintores,**
shown with additional Santa Fe
artists at the Los Angeles
County Museum, 1923. Died
in Santa Fe, 1977.

Public Collections: Brooklyn
Museum; California Palace of the
Legion of Honor, San Francisco;
Delaware Art Center,
Wilmington; University of
Oklahoma Museum of Art,
Norman; Whitney Museum of
American Art, New York.

GUSTAVE BAUMANN

**WINTER CEREMONY —
DEER DANCE,** 1922,
oil on panel
30-7/16″ x 52-9/16″
(77.3cm. x 133.4cm.)
Gift of Florence Dibell
Bartlett, 1948

Born in Magdeburg, Germany,
1881. Grew up in Chicago;
studied at Kunstgewerbe Schule,
Munich and the Art Institute of
Chicago. Settled in Santa Fe in
1918, worked as a painter,
printmaker and carver of
marionettes. Baumann was
internationally famous as the
creator of color woodcuts
depicting a wide range of
Southwestern subjects. In 1952
was made a Fellow of the School
of American Research. Died in
Santa Fe, 1971.

Public Collections: Art Institute
of Chicago; Sheldon Memorial
Art Gallery, University of
Nebraska-Lincoln.

GEORGE BELLOWS

CHIMAYO, 1917,
oil on canvas
30-7/16″ x 44-1/4″
(77.3cm. x 112.3cm.)
Gift of an Anonymous
Donor

Born in Columbus, Ohio, 1882. Studied at Ohio State University; New York with Robert Henri, Kenneth Hayes Miller and H. G. Maratta. Taught at the Art Students League, New York; Ferrer School; and the Art Institute of Chicago. Elected to the National Academy of Design, 1913. Helped organize the Armory Show. Bellows' stay in New Mexico was brief, one month in the summer of 1917, but from that experience he created a number of notable works. Died in New York, 1925.

Solo Exhibitions: Art Institute of Chicago, 1946; Museum of New Mexico, Santa Fe, 1952; National Gallery of Art, Washington, D.C., 1957; Gallery of Modern Art including the Huntington Hartford Collection, New York, 1966.

Public Collections:
Albright-Knox Art Gallery, Buffalo; Art Institute of Chicago; Brooklyn Museum; Cleveland Museum of Art; Columbus Museum of Art; Corcoran Gallery of Art, Washington, D.C.; Des Moines Art Center; Los Angeles County Museum; Metropolitan Museum of Art, New York; Museum of Fine Arts, Boston; New York Public Library; Roswell Museum and Art Center; Sheldon Memorial Art Gallery, University of Nebraska-Lincoln; Whitney Museum of American Art, New York; Worcester Art Museum.

TOM BENRIMO

MORADA, 1940,
gouache on paper
6-5/8" x 11-3/8"
(17cm. x 29cm.)
Gift of Rebecca Salsbury
James

Born in San Francisco, 1887.
Largely self-taught. Studied
briefly at Art Students League.
Taught at Pratt Institute, New
York, 1935-39. Early career
devoted to commercial art and
theater design. Moved to Taos,
1939. Died there, 1958.

Solo Exhibitions: Witte
Museum, San Antonio, 1953;
San Francisco Museum of Art,
1954; Museum of Modern Art,
New York, 1955; Ft. Worth Art
Center, 1958; Museum of New
Mexico, Santa Fe, 1958 and
1973; Ft. Worth Art Center,
1965 (shown in Oklahoma City,
Roswell, El Paso and Dallas).

Public Collections: Cincinnati
Art Museum; Dallas Museum of
Fine Arts; Harwood Foundation
of the University of New Mexico,
Taos; Krannert Art Museum,
University of Illinois,
Champaign-Urbana; Sheldon
Memorial Art Gallery,
University of Nebraska-Lincoln;
Whitney Museum of American
Art, New York.

OSCAR E. BERNINGHAUS

THE RABBIT HUNTER,
oil on canvas
34-7/16″ x 39-1/2″
(87.5cm x 100.3cm.)
Gift of Mr. and Mrs. John
A. Hill, 1975

CORN DANCE DAY, TAOS,
oil on canvas
20″ x 24″
(51cm. x 61cm.)
Purchase Award, 1949

Born in St. Louis, Missouri,
1874. Studied at St. Louis School
of Fine Arts; Washington
University, St. Louis where he
also taught. Well known as an
illustrator and lithographer
before moving to New Mexico.
One of the six original members
of the Taos Society of Artists,
organized in 1912. Established
full time residence in Taos,
1925. Died there 1952.

Public Collections: Anschutz
Collection, Denver; Arkansas Art
Center, Little Rock; Eiteljorg
Collection, Indianapolis; Thomas
Gilcrease Institute of American
History and Art, Tulsa; Harmsen
Collection, Denver; Harwood
Foundation of the University of
New Mexico, Taos; Philbrook Art
Center, Tulsa; St. Louis Art
Museum.

EMIL BISTTRAM

HOPI SNAKE DANCERS,
1933,
oil on canvas
41-3/16" x 45"
(104cm. x 114.3cm.)
Gift of the artist, 1966

Born in Hungary, 1895. Studied
at National Academy of Design;
Cooper Union; and the Art
Students League, New York,
with Howard Giles and Jay
Hambidge. Taught at New York
School of Fine and Applied Arts;
Master Institute of Roerich
Museum, New York; Director of
the Bisttram School of Fine Arts,
Taos. Died in Taos, 1976.

Public Collections: Harwood
Foundation of the University of
New Mexico, Taos; Jonson
Collection, University of New
Mexico, Albuquerque.

ERNEST L. BLUMENSCHEIN

DANCE AT TAOS, 1923,
oil on canvas
24″ x 27″
(61cm. x 68.5cm.)
Gift of Florence Dibell
Bartlett

UNTITLED (Adobe Church),
oil on canvas,
16″ x 20″
(40.5cm. x 50.8cm.)
Gift of Helen Greene, 1964

Born in Pittsburgh,
Pennsylvania, 1874. Studied at
Cincinnati Art Academy;
Academie Julien and Ecole des
Beaux Arts, Paris. Taught at the
Art Students League, New York.
In 1927, elected to National
Academy of Design; Honorary
Master of Arts, University of
New Mexico, 1947; named
Honorary Fellow in Fine Arts by
the School of American Research,
Museum of New Mexico, Santa
Fe, 1948. Died in Taos, 1960.

Solo Exhibitions: Grand Central
Art Galleries, New York, 1927;
Museum of New Mexico, Santa
Fe, 1948 and 1958; Colorado
Springs Fine Arts Center, 1978.

Public Collections: Art Gallery
of Toronto; Cincinnati Art
Museum; Dayton Art Institute;
Ft. Worth Art Center; Thomas
Gilcrease Institute of American
History and Art, Tulsa; Harwood
Foundation of the University of
New Mexico, Taos; Los Angeles
County Museum of Art; Museum
of Modern Art, New York;
University of New Mexico,
Albuquerque.

DOROTHY E. BRETT

INDIAN MAN (Taos Pueblo Indian), 1937,
oil on canvas board
23-1/2″ x 17-7/8″
(59.7cm. x 45.4cm.)
Bequest of Vivian Sloan Fiske, 1979

THE FIRST BORN, 1937,
oil on canvas
50-1/4″ x 24-1/16″
(172cm. x 61.1cm.)
Gift of the Millicent Rogers Foundation

Born in London, England, 1883. Studied at Slade School of Art, London. Came to New Mexico with D. H. Lawrence in 1924, remaining a resident of Taos until her death. Although initially trained as a portrait painter, once in New Mexico she became fascinated with the Indians and their ceremonials. Died in Taos, 1977.

Public Collections: Harwood Foundation of the University of New Mexico, Taos; Millicent A. Rogers Memorial Museum, Taos; Texas Tech University, Lubbock.

36

FRANK CHEATHAM

PERIODIC STRUCTURE,
acrylic on canvas
45-1/16″ x 50″
(114.4cm. x 127cm.)
Purchase Award, Museum
of New Mexico Foundation,
1974

Born in Beeville, Texas, 1936.
Studied at Chouinard Art
Institute; Otis Art Institute, Los
Angeles, with Lorser Feitelson,
Louis Danziger and Arthur
Ames. Received Certificate of
Merit, Art Directors Club, New
York, 1960, 1970 and 1974;
Award of Excellence, American
Institute of Graphic Arts, 1963,
1965, 1972 and 1974. Presently
Associate Professor of Design,
Texas Tech University, Lubbock.

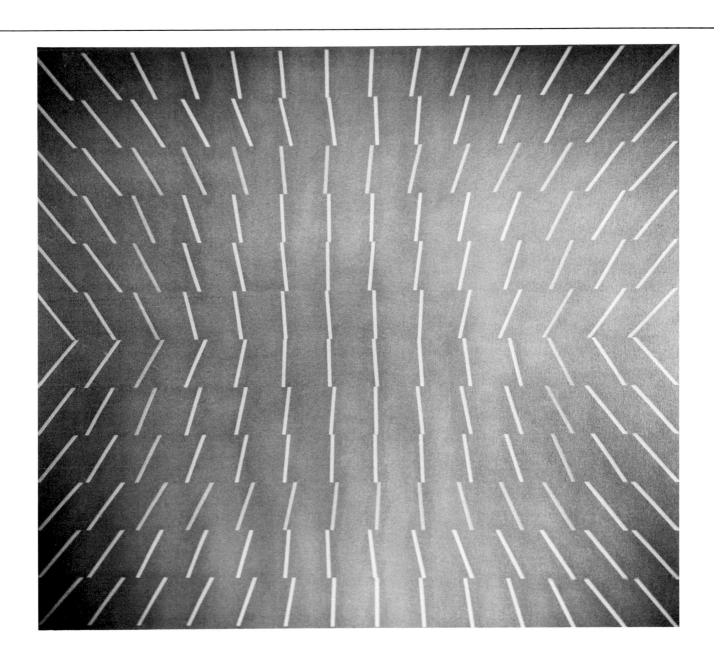

ANDREW DASBURG

PORTRAIT OF CHARLES AUGUSTUS FICKE,
oil on canvas
50-1/4″ x 36-1/8″
(127.5cm. x 91.8cm.)
Gift of Stanhope Blunt Ficke

Born in Paris, France, 1887. Came to United States, 1892. Studied at the Art Students League, New York with Kenyon Cox, Birge Harrison and at the New York School of Arts with Robert Henri. Taught at the Art Students League. To Taos in 1918 at the invitation of Mable Dodge Luhan. Permanent resident, 1933. Awarded a Guggenheim Fellowship, 1932; Honorary Doctor of Fine Arts degree from the University of New Mexico, 1958. Died in Taos, 1979.

Solo Exhibitions: Dallas Museum of Fine Arts, 1957; American Federation of Arts, New York, 1959 (traveling exhibition); Taos Art Association, 1966; Governor's Gallery, Santa Fe, 1976; Art Museum, University of New Mexico, Alubquerque (traveling exhibition), 1979.

Public Collections: Cincinnati Art Museum; Dallas Museum of Fine Arts; Denver Art Museum; Harwood Foundation of the University of New Mexico, Taos; Los Angeles County Museum of Art; Metropolitan Museum of Art, New York; Nelson Gallery of Art, Kansas City; San Francisco Museum of Art; Santa Barbara Museum of Art; Sheldon Memorial Art Gallery, University of Nebraska-Lincoln; Whitney Museum of American Art, New York.

RANDALL DAVEY

SPRING IN SANTA FE,
oil on masonite
21-3/8" x 25-9/16"
(55.3cm. x 65.5cm.)
Gift of Mr. and Mrs. Earl C.
Kaufman

**WINTER LANDSCAPE —
    NEW MEXICO,** 1923,
oil on canvas
26" x 32"
(66cm. x 81.2cm.)
Gift of Mrs. E. G. Cullum
and the Museum of New
Mexico Foundation

Born in East Orange, New
Jersey, 1887. Studied at Cornell
University (architecture); Henri
School of Painting and Arts
Students League, New York.
Taught at Chicago Art Institute;
Broadmoor Art Academy,
Colorado Springs and the
University of New Mexico. To
Santa Fe with John Sloan, 1919.
Elected to the National Academy
of Design, 1938; named an
Honorary Fellow in Fine Arts by
the School of American
Research, Museum of New
Mexico, Santa Fe, 1957. Died in
Santa Fe, 1964.

Solo Exhibitions: Museum of
New Mexico, Santa Fe, 1957;
Roswell Museum and Art
Center, 1963.

Public Collections: Cleveland
Museum of Art; Corcoran
Gallery of Art, Washington,
D.C.; Delaware Art Center,
Wilmington; Detroit Institute of
Arts; Whitney Museum of
American Art, New York.

FREMONT F. ELLIS

ADIOS AMIGO — HASTA LA VISTA, 1969,
oil on masonite
25" x 29"
(63.5cm. x 76cm.)
Gift of the artist

WHITE HOUSE RUINS,
oil on canvas
24" x 30-1/2"
(61cm. x 77.5cm.)
Gift from Marjorie D.
Myers, 1971

Born in Virginia City, Montana, 1897. Studied briefly at Art Students League in New York, 1925, but is primarily a self-taught artist. He arrived in Santa Fe in 1919 and shortly after his arrival joined with Willard Nash, Josef Bakos, Walter Mruk and Will Shuster to form the group known as **Los Cinco Pintores.** Lives in Santa Fe.

Public Collections: El Paso Museum of Art; Thomas Gilcrease Institute of American History and Art, Tulsa; Texas Tech University, Lubbock; University of California, Los Angeles.

R. C. GORMAN

HOMAGE TO SPIDERWOMAN,
acrylic
60″ x 50″
(152.4cm. x 127cm.)
Gift of the artist

Born in Chinle, Arizona, 1933. Studied at Northern Arizona University with Jack Salter and Ellery Gibson and at Mexico City College with Carlos Merida. Received awards at the American Indian Art Exhibit, Oakland, 1966; First Scottsdale National Indian Exhibit, 1967; and Heard Museum Guild Exhibit, Phoenix, 1968.

Solo Exhibition: Museum of the American Indian, Heye Foundation, New York, 1975.

Public Collections: Gonzaga University, Spokane; Heard Museum, Phoenix; Metropolitan Museum of Art, New York; Museum of the American Indian, Heye Foundation, New York; Museum of Indian Arts, San Francisco; Museum of Northern Arizona, Flagstaff; Philbrook Art Center, Tulsa.

MARSDEN HARTLEY

STILL LIFE, oil,
27″ x 22″
(68.6cm. x 55.9cm.)
Gift of Rebecca Salsbury
James

Born in Lewiston, Maine, 1877.
Studied at Cleveland School of
Art; Chase School; and the
National Academy of Design,
New York. Hartley spent some
months in New Mexico in 1918
and again in 1919. A series of
works entitled "New Mexico
Recollections" was executed in
1923 and 1924. Died in
Ellsworth, Maine, 1943.

Solo Exhibitions: American
Federation of Arts, New York,
1960 (traveling exhibition);
University Galleries, University
of Southern California, Los
Angeles, 1968; Smithsonian
Institution Traveling Exhibition
Service, 1970; University
Gallery, University of
Minnesota, 1979; Whitney
Museum of American Art, New
York, 1980.

Public Collections: Art Institute
of Chicago; Columbus Museum
of Art; Fogg Art Museum,
Harvard University, Cambridge;
Metropolitan Museum of Art,
New York; Museum of Modern
Art, New York; National Gallery
of Art, Washington, D.C.; Nelson
Gallery of Art, Kansas City;
Roswell Museum and Art Center;
Sheldon Memorial Art Gallery,
University of Nebraska-Lincoln;
University Gallery, University of
Minnesota, Minneapolis; Walker
Art Center, Minneapolis;
Whitney Museum of American
Art.

GREEN CORN DANCE,
oil
18″ x 24″
(46cm. x 61cm.)
Gift of Mrs. Richard M.
Delafield

Born in Medford, Massachusetts, 1877. Studied at Massachuesetts Normal Art School; Boston Museum of Fine Arts School with Edmund C. Tarball. Taught at Chicago Academy of Fine Arts. Produced murals for Frank Lloyd Wright's Midway Gardens in Chicago. Moved to Santa Fe, 1916, where he was active as an architect, working on the restoration of Sena Plaza and designing the Museum of Navajo Ceremonial Art. His work as a painter was centered on the depiction of the dance rituals of the surrounding pueblos. Died in Santa Fe, 1943.

Solo Exhibition: National Collection of Fine Arts, Smithsonian Institution, Washington, D.C., 1978.

Public Collections: Art Institute of Chicago; Arthur Johnson Memorial Library, Raton, NM.

E. MARTIN HENNINGS

THE RENDEZVOUS,
oil on canvas
25-1/4″ x 30″
(64cm. x 76.2cm.)
Purchased by Museum of
New Mexico Foundation,
1977

Born in Pennsgrove, New
Jersey, 1886. Studied at the Art
Institute of Chicago; Munich
Academy with Walter Thor; and
Royal Academy, Munich with
Angelo Junk. Was elected to
membership in the Taos Society
of Artists, 1921. Died in Taos,
1956.

Public Collections: Eiteljorg
Collection, Indianapolis; Thomas
Gilcrease Institute of American
History and Art, Tulsa; Harmsen
Collection, Denver; Los Angeles
County Museum of Art; Museum
of Fine Arts, Houston;
Pennsylvania Academy of the
Fine Arts, Philadelphia.

ROBERT HENRI

DIEGUITO, DRUMMER OF THE EAGLE DANCE, SAN ILDEFONSO, 1916, oil
65-3/8" x 40-7/16"
(166cm. x 102.7cm.)
Gift of the artist, 1916

Born in Cincinnati, Ohio, 1865. Studied at Pennsylvania Academy of Fine Arts; École des Beaux-Arts; and Academie Julien, Paris, with Adolphe Bouguereau and Robert Fleury. Taught at Chase School in New York and Art Students League. Founded own school in New York, 1909. First visited Santa Fe 1916, spent six months there in 1922, last visit in 1925. Died, New York, 1929.

Solo Exhibitions: Metropolitan Museum of Art, New York, 1931; Sheldon Memorial Art Gallery, University of Nebraska-Lincoln, 1965 and 1971; New York Cultural Center, 1969.

Public Collections: Corcoran Gallery of Art, Washington, D.C.; Indianapolis Museum of Art; Los Angeles County Museum of Art; Metropolitan Museum of Art, New York; Sheldon Memorial Art Gallery, University of Nebraska-Lincoln; Whitney Museum of American Art, New York; Wichita Art Museum.

VICTOR HIGGINS

**MERCEDES' FIRST
 COMMUNION,**
oil on canvas
18-7/8″ x 27″
(47.9cm. x 68.5cm.)
Gift of Mr. and Mrs. Wally
Sargent

WINGED VICTORY,
oil on canvas
40″ x 43-1/4″
(101.8cm. x 109.5cm.)
Gift of Mrs. Joan Higgins
Reed

**LANDSCAPE TAOS,
 NEW MEXICO,**
watercolor
(sight) 22″ x 15″
(56.6cm. x 38.2cm.)
Gift from Laura Hersloff,
1980

Born in Shelbyville, Indiana,
1884. Studied at Art Institute of
Chicago; Chicago Academy of
Fine Arts; in Paris with René
Menard and Lucien Simon; in
Munich with Haas von Hyeck;
taught at Chicago Academy of
Fine Arts. To New Mexico in
1914. Member of the Taos
Society of Artists. Died Taos,
1949.

Solo Exhibitions: Museum of
New Mexico, Santa Fe, 1971;
University of New Mexico,
Albuquerque, 1972; Phoenix Art
Museum, 1973; Art Gallery of
the University of Notre Dame,
South Bend, 1975.

Public Collections: Adkins
Collection, Tulsa; Art Institute of
Chicago; Butler Institute of
American Art, Youngstown;
Thomas Gilcrease Institute of
American History and Art,
Tulsa; Harwood Foundation of
the University of New Mexico,
Taos; Los Angeles County
Museum of Art; Pennsylvania
Academy of the Fine Arts,
Philadelphia.

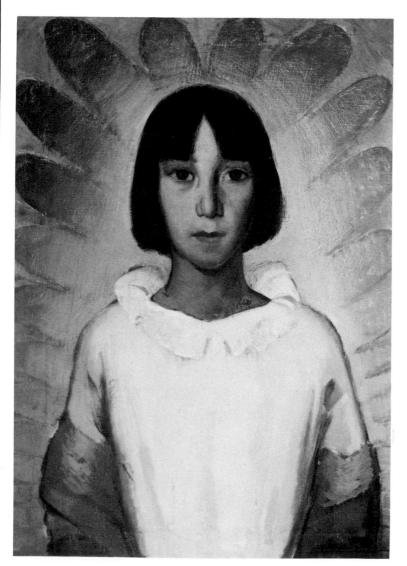

PETER HURD

HIGHWAY AT DUSK, 1955, egg tempera on masonite 31-3/4″ x 47-7/8″ (80.8cm. x 121.9cm.) Gift of the artist and private contributors

Born in Roswell, New Mexico, 1904. Studied at United States Military Academy; Haverford College; and Pennsylvania Academy of the Fine Arts with N. C. Wyeth. Member, National Academy of Design. Lives near Roswell.

Solo Exhibitions: Amon Carter Museum of Western Art, Ft. Worth, 1964; California Palace of the Legion of Honor, San Francisco, 1965.

Public Collections: Dallas Art Museum; Delaware Art Center, Wilmington; Metropolitan Museum of Art, New York; National Gallery, Edinburgh, Scotland; National Portrait Gallery, Smithsonian Institution, Washington, D.C.; Nelson Gallery of Art, Kansas City; Roswell Museum and Art Center.

DOUGLAS JOHNSON

**THE BUTTERFLY PEOPLE
RETURN TO THE
THIRD WORLD,** 1971,
casein on paper
(sight) 17-3/8" x 21-1/8"
(44cm. x 53.6cm.)
Gift of Mr. and Mrs. Philip
Casady

Born in Portland, Oregon, 1946.
Self taught as an artist. Has
exhibited primarily in New
Mexico. A retrospective
exhibition (1969-1975) was
presented by the Jamison
Gallery, Santa Fe in 1975. Lives
in Abiquiu.

RAYMOND JONSON

WATERCOLOR NO. 22,
  1939,
watercolor on paper
(sight) 17-7/8″ x 20″
(70.7cm. x 50.8cm.)
Anonymous Donor

Born in Chariton, Iowa, 1891.
Studied at Portland Museum Art
School; Chicago Academy of Fine
Arts with B.J.O. Nordfeldt; and
the Art Institute of Chicago.
Taught at Chicago Academy of
Fine Arts and University of New
Mexico. The Jonson Gallery at
the University of New Mexico,
Albuquerque, opened with a
Jonson retrospective in 1950.
Made an Honorary Fellow of the
School of American Research,
Santa Fe, 1951. Retired from
teaching in 1954, remains as
Director of the Jonson Gallery.
Received the honorary degree of
Doctor of Humane Letters from
the University of New Mexico,
1971. Lives in Albuquerque.

Solo Exhibition: Museum of
New Mexico, Santa Fe, 1956.

Public Collections: Cincinnati
Art Museum; Dallas Museum of
Fine Arts; Nelson Gallery of Art,
Kansas City; Rose Art Museum,
Brandeis University, Waltham,
MA; Sheldon Memorial Art
Gallery, University of
Nebraska-Lincoln.

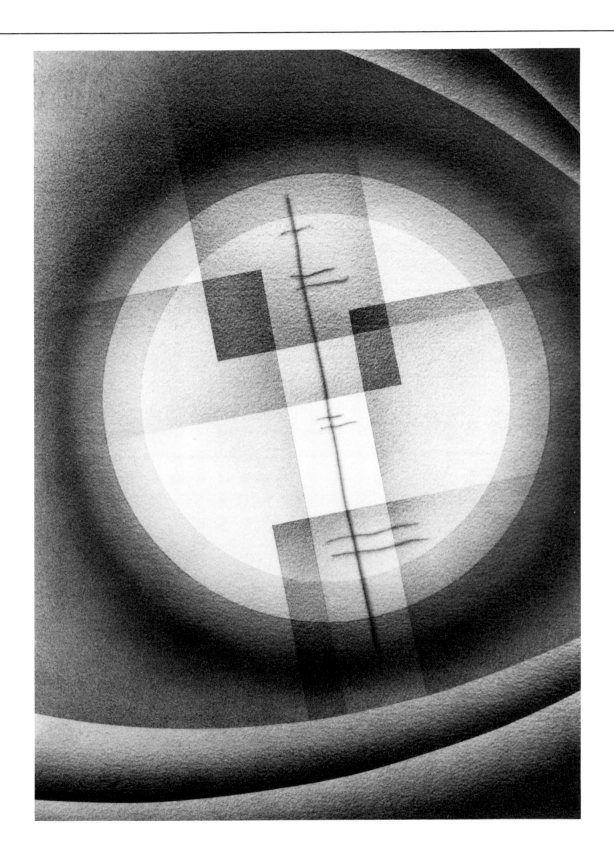

LEON KROLL

SANTA FE HILLS, 1917,
oil on canvas
34" x 40-1/4"
(86.3cm. x 101.9cm.)
Gift of the Museum of New
Mexico Foundation

Born in New York, 1884.
Studied at the Art Students
League with John Twachtman;
National Academy of Design;
and in Paris with Jean Paul
Laurens. Spent part of the
summer of 1917 in Santa Fe in
the company of George Bellows
and Robert Henri. Taught at the
Art Institute in Chicago. Elected
to the National Academy of
Design, 1927. Died in New York,
1974.

Public Collections: Art Institute
of Chicago; Cleveland Museum of
Art; Dayton Art Institute;
Metropolitan Museum of Art,
New York; Pennsylvania
Academy of the Fine Arts,
Philadelphia; Sheldon Memorial
Art Gallery, University of
Nebraska-Lincoln; St. Louis Art
Museum; Whitney Museum of
American Art, New York.

LEE MULLICAN

**MEDITATION ON LEAVES
  IN A POND,** 1962,
oil on canvas
40-1/8" x 75-3/16"
(101.7cm. x 190.9cm.)
Gift of Mr. and Mrs. Gifford
Phillips, 1980

Born in Chickasha, Oklahoma,
1919. Studied at Abilene
Christian College; University of
Oklahoma; Kansas City Art
Institute with Fletcher Martin;
and San Francisco Art Institute
with Stanley Hayter. Taught at
University of California, Los
Angeles, 1962-78. Lives in Santa
Monica, California and Taos.

Solo Exhibitions: San Francisco
Museum of Art, 1949 and 1965;
Oklahoma Art Center, Oklahoma
City, 1951 and 1967; Museo
Nacional de Bellas Artes,
Santiago, Chile, 1958; Santa
Barbara Museum of Art, 1958
and 1973; University of
California, Los Angeles, 1959;
Pasadena Museum of Art, 1961.

Public Collections: Museum of
Modern Art, New York;
Oklahoma Art Center, Oklahoma
City; Phillips Collection,
Washington, D.C.; San Francisco
Museum of Art; Santa Barbara
Museum of Art.

WILLARD NASH

**PORTRAIT OF
  EDITH NASH,**
oil on canvas
36-1/4" x 24"
(92cm. x 61cm.)
Gift of Emma B. Mosley

Born in Philadelphia,
Pennsylvania, 1898. Studied in
Detroit with John P. Wicker.
Joined four other young artists
in forming **Los Cinco Pintores,**
1921. He taught at Broadmoor
Art Academy, Colorado Springs;
San Francisco Art School; and
the Art Center School, Los
Angeles. Died in Los Angeles,
1943.

Public Collections: Jonson
Collection, University of New
Mexico, Albuquerque; Los
Angeles County Museum of Art;
Lovelace Foundation for Medical
Education and Research,
Albuquerque.

B.J.O. NORDFELDT

ANTELOPE DANCE, 1919, oil on canvas
35-5/8" x 43"
(90.5cm. x 109.2cm.)
Gift of the Archeological Society, Purchase Fund of the Museum of New Mexico and Friends of Southwestern Art.

Born in Sweden, 1878. To the United States, 1891. Studied at the Art Institute of Chicago; with Albert Herter in New York; Jean Paul Laurens in Paris; and with Frank Fletcher in England. To Santa Fe, 1919, where he maintained a home until 1937. From 1900 through the Twenties Nordfeldt was continuously active as a printmaker, his woodcuts and etchings being widely shown. Died in Henderson, Texas, 1955.

Solo Exhibitions: Daniel Gallery, New York, 1913 and 1917; National Arts Club, New York, 1917; Arts Club of Chicago, 1920; Denver Art Museum, 1929; Minneapolis Institute of Arts, 1933; Lilienfeld Gallery, New York, 1937; Hudson Walker Gallery, New York, 1940; Passedoit Gallery, New York, 1944 and 1955.

Public Collections: Corcoran Gallery of Art, Washington, D.C.; Denver Art Museum; Joseph H. Hirshhorn Museum and Sculpture Garden, Smithsonian Institution, Washington, D.C.; Jonson Collection, University of New Mexico, Albuquerque; Metropolitan Museum of Art, New York; Minneapolis Institute of Arts; Sheldon Memorial Art Gallery, University of Nebraska-Lincoln; University of Oklahoma Museum of Art, Norman; University of Oregon, Eugene; Worcester Art Museum.

MARCIA LEE OLIVER

JUNE SERIES #1, 1969,
acrylic on canvas
59-3/4″ x 59-7/8″
(151.8cm. x 152cm.)
Gift of the artist

Studied at University of
Alabama; University of Illinois;
Provincetown School of Fine
Arts; New York University; and
San Jose State College. Received
Wurlitzer Foundation Residency
Award, Taos, New Mexico,
1968-69. Lives in Taos.

Solo Exhibitions: Cellini
Gallery, San Francisco, 1968;
Pensacola Art Center, 1969;
Stables Gallery, Taos, 1978;
Colorado Mountain College,
Leadville, 1979.

Public Collections: Pensacola
Art Center; San Jose State
College; University of Arizona,
Tucson; University of Illinois.

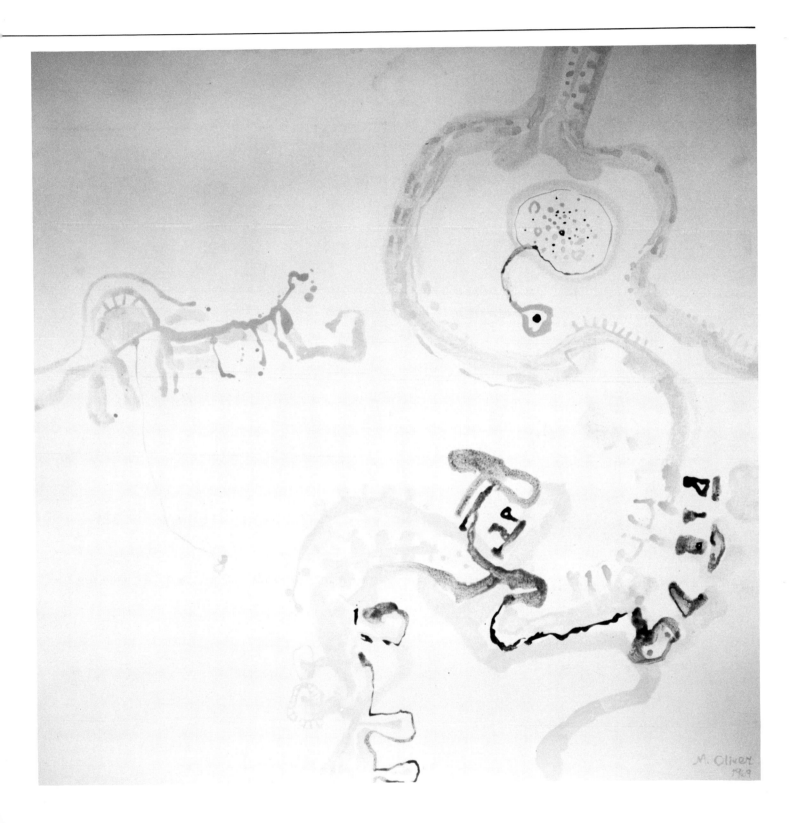

BERT G. PHILLIPS

OUR WASHERWOMAN'S FAMILY-NEW MEXICO (La Lavanderia), c. 1918, oil on canvas
40-1/2″ x 41-5/8″
(102.8cm. x 105.7cm.)
Gift of Governor and Mrs. Arthur Seligman

Born in Hudson, New York, 1868. Studied at National Academy of Design; Art Students League in New York; and Académie Julien, Paris with Benjamin Constant and Jean Paul Laurens. At the urging of Joseph H. Sharp, Phillips and Ernest Blumenschein arrived in Taos in 1893 and Phillips was the first artist to establish a studio there. One of the six original members of the Taos Society of Artists organized in 1912. Died in San Diego, California, 1956.

Solo Exhibition: Babcock Galleries, New York, 1920.

Public Collections: Anschutz Collection, Denver; Eiteljorg Collection, Indianapolis; Thomas Gilcrease Institute of American History and Art, Tulsa; Philbrook Art Center, Tulsa; Woolaroc Museum, Bartlesville, OK.

FRITZ SCHOLDER

SUPER INDIAN,
oil
102″ x 68″
(305.3cm. x 172.7cm.)
Gift of the artist

Born in Breckenridge, Minnesota, 1937. While in high school studied with Oscar Howe. Studied at the University of Kansas, Lawrence; Wisconsin State University, Superior; Sacramento City College with Wayne Thiebaud; Sacramento State University; and University of Arizona, Tucson. Taught at Institute of American Indian Arts, Santa Fe, 1968-69. Artist-in-residence, Dartmouth College, 1973. Lives in Galisteo, New Mexico and Scottsdale, Arizona.

Solo Exhibitions: Sacramento State University Art Gallery, 1958; E. B. Crocker Art Gallery, Sacramento, California, 1972; Cedar Rapids Art Center, 1972; Sheldon Memorial Art Gallery, University of Nebraska-Lincoln, 1973; Massachusetts Institute of Technology, 1973; Jaffe-Friede Gallery, Dartmouth College, Hanover, New Hampshire, 1973; Brooks Memorial Art Gallery, Memphis, 1974; Wheelwright Museum, Santa Fe, 1977; Oakland Art Museum, 1977; Saginaw Art Museum, Michigan, 1977.

Public Collections: Brooklyn Museum; Dallas Museum of Fine Arts; Houston Museum of Fine Arts; Milwaukee Art Center; Phoenix Art Museum; San Diego Gallery of Fine Art; Sheldon Memorial Art Gallery, University of Nebraska-Lincoln.

ELMER SCHOOLEY

MIXED CONIFERS,
oil on canvas
80-1/16″ x 90″
(203.3cm. x 228.5cm.)
Gift of the Museum of New
Mexico Foundation

Born in Lawrence, Kansas,
1916. Studied at University of
Colorado and Iowa State
University. Has taught at
Western New Mexico University,
Silver City; presently Professor
of Arts and Crafts at New
Mexico Highlands University,
Las Vegas; Artist-in-Residence,
Roswell Museum, New Mexico,
1977-78. Lives in Montezuma,
New Mexico.

Public Collections: Metropolitan
Museum of Art, New York;
Museum of Modern Art, New
York; Roswell Museum and Art
Center, Roswell.

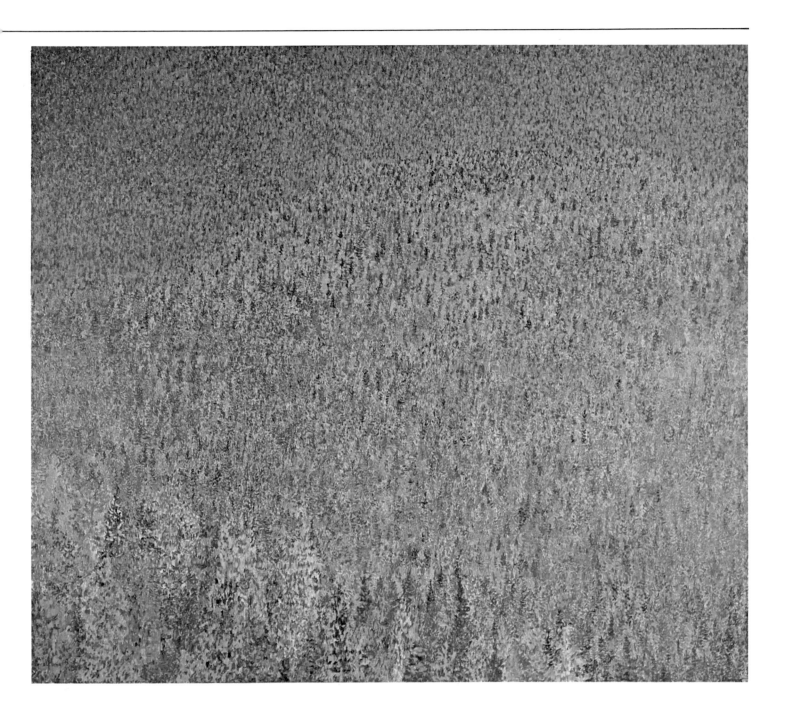

JOSEPH H. SHARP

TAOS INDIAN, 1911,
oil on canvas
24″ x 20″
(61cm. x 50.8cm.)
Gift of the artist

TAOS INDIAN PORTRAIT,
   1914,
oil on canvas
30″ x 24″
(76.2cm. x 61cm.)
Gift of the artist

Born in Bridgeport, Ohio, 1859.
Studied at McMicken School of
Design, Cincinnati; Antwerp
with Charles Verlet; Munich
with Carl Marr; and in Paris
with Jean Paul Laurens and
Benjamin Constant. Taught at
Cincinnati Academy of Art. One
of the founding members of the
Taos Society of Artists. Sharp
was particularly admired for the
accuracy of his depiction of the
physical characteristics of the
various tribes as well as for the
documentation of costumes,
accoutrements and rituals. Died
in Pasadena, California, 1953.

Public Collections: Butler
Institute of American Art,
Youngstown; Cincinnati Art
Museum; Thomas Gilcrease
Institute of American History
and Art, Tulsa; National Cowboy
Hall of Fame, Oklahoma City;
Southwest Museum, Los
Angeles; St. Louis Art Museum;
Department of Anthropology,
University of California,
Berkeley; Woolaroc Museum,
Bartlesville, OK.

JOHN SLOAN

**UNDER THE OLD
  PORTAL,** 1919,
oil on canvas
24″ x 20″
(61cm. x 51cm.)
Gift of Julius Gans

ANCESTRAL SPIRITS,
1919,
oil on canvas
24″ x 20″
(61cm. x 50.8cm.)
Gift of Dr. Edgar L. Hewett

Born in Lock Haven,
Pennsylvania, 1871. Studied at
Pennsylvania Academy of Fine
Arts with Thomas Anshutz and
with Robert Henri. Taught at
Art Students League and George
Luks Art School, New York.
Arrived in Santa Fe, 1919,
maintaining a residence there
for the rest of his life, dividing
his time between New York and
the Southwest. Died in Hanover,
New Hampshire, 1951.

Solo Exhibitions: Museum of
New Mexico, Santa Fe, 1941;
Dartmouth College, Hanover,
1946; Whitney Museum of
American Art, New York, 1952;
National Gallery of Art,
Washington, D.C., 1971;
Museum of New Mexico, Santa
Fe, 1971.

Public Collection: Anschutz
Collection, Denver; Bowdoin
College Museum of Fine Arts,
Brunswick, ME; Colorado
Springs Fine Arts Center;
Thomas Gilcrease Institute of
American History and Art,
Tulsa; Mint Museum of Art,
Charlotte; Newark Museum;
Parrish Art Museum,
Southhampton, NY; Sheldon
Memorial Art Gallery,
University of Nebraska-Lincoln.

EARL STROH

SAND STORM, 1951
oil on masonite
8-15/16" x 19-7/8"
(22cm. x 50.4cm.)
Gift of the Helene Wurlitzer
Foundation, Taos

Born in Buffalo, New York, 1924. Studied at the Art Institute of Buffalo; the Art Students League, New York with Edwin Dickinson; the University of New Mexico; in Taos with Andrew Dasburg and Tom Benrimo; and with Jonny Friedlander in Paris. Taught painting at the Art Institute of Buffalo, 1943-45. Has exhibited widely as a painter and printmaker. Lives in Taos.

Solo Exhibitions: Museum of New Mexico, Santa Fe, 1948, 1964 and 1979; Jonson Gallery, University of New Mexico, Albuquerque, 1953 and 1963; Ft. Worth Art Center, 1961 and 1965; Oklahoma Art Center, Oklahoma City, 1961; Birger Sandzen Memorial Gallery, Lindsborg, KS, 1966; Roswell Museum and Art Center, 1967.

Public Collections: Art Institute of Chicago; Cincinnati Art Museum; Dallas Museum of Fine Arts; Denver Art Museum; Ft. Worth Art Center; Harwood Foundation of the University of New Mexico, Taos; Helene Wurlitzer Foundation, Taos; Jonson Gallery, University of New Mexico, Albuquerque; Metropolitan Museum of Art, New York; University of New Mexico, Albuquerque; New Mexico Highlands University, Las Vegas.

WALTER UFER

WILD PLUM BLOSSOMS,
1922,
oil on canvas
10-1/2″ x 12-3/8″
(26.7cm. x 31.4cm.)
Gift of Florence Dibell
Bartlett

CHANCE ENCOUNTER,
oil on canvas
20″ x 25″
(50.8cm. x 63.5cm.)
Gift of Museum of New
Mexico Foundation in honor
and memory of Mr. and
Mrs. Lewis Barker, 1980

Born in Louisville, Kentucky,
1876. Studied at Royal
Academy, Dresden; Munich
Academy with Walter Thor; and
the Art Institute of Chicago. To
Taos in 1914, elected a member
of the Taos Society of Artists.
Winner of third prize at the
Carnegie International
Exhibition, 1920. Died in
Albuquerque, 1936.

Solo Exhibition: Phoenix Art
Museum, 1970.

Public Collections: Anschutz
Collection, Denver; Art Institute
of Chicago; Baltimore Museum
of Art; Corcoran Gallery of Art,
Washington, D.C.; Eiteljorg
Collection, Indianapolis; Thomas
Gilcrease Institute of American
History of Art, Tulsa; Los
Angeles County Museum of Art;
University of Notre Dame, South
Bend.

THEODORE VAN SOELEN

THE PORTUGUESE,
oil on canvas
36″ x 30″
(91.5cm. x 76.2cm.)
Lent by Don Van Soelen

A SANTA FE HILLSIDE,
  c. 1924,
oil on canvas
34″ x 36″
(86.3cm. x 91.5cm.)
Gift of Henry Dendahl in
memory of Johanna S.
Dendahl

Born in St. Paul, Minnesota,
1890. Studied at St. Paul
Institute of Arts and Sciences
and Pennsylvania Academy of
the Fine Arts. Elected to
National Academy of Design,
1940; elected Honorary Fellow in
Fine Arts by the School of
American Research, Santa Fe,
1960. Died in Tesuque, 1964.

Solo Exhibitions: Denver Art
Museum, 1931; Museum of New
Mexico, Santa Fe, 1960.

Public Collections: Anschutz
Collection, Denver; Eiteljorg
Collection, Indianapolis;
Everhart Museum, Scranton,
PA; Harmsen Collection, Denver;
Loomis Institute, Windsor, CT;
Pennsylvania Academy of the
Fine Arts, Philadelphia; Texas
Tech University, Lubbock.

CADY WELLS

UNTITLED (New Mexico Landscape),
watercolor
(sight) 10-1/2″ x 14-1/2″
(26.7cm. x 36.9cm.)
Bequest of Vivian Sloan Fiske

Born in Southbridge, Massachusetts, 1904. Initially studied music and theater design, then art history at Harvard University; art at the University of Arizona; and with Andrew Dasburg in Taos. Died in Santa Fe, 1954.

Solo Exhibitions: School of American Research, Santa Fe, 1956; University of New Mexico Art Museum, 1967.

Public Collections: Addison Gallery of American Art, Andover; California Palace of the Legion of Honor, San Francisco; Fine Arts Gallery of San Diego; Fogg Art Museum, Harvard University, Cambridge; Jonson Collection, University of New Mexico, Albuquerque; Wadsworth Athenaeum, Hartford.

JOHN WENGER

**ANOTHER COWBOY
    PAINTING,** 1975,
oil on canvas
71-3/4″ x 65-7/8″
(182.1cm. x 167.4cm.)
Purchased with the aid of
funds from the National
Endowment for the Arts,
1975

Born in Salem, Oregon, 1940.
Studied at the Universities of
Colorado and Arizona. Taught at
the University of Arizona;
Tucson Art Center and is
presently Assistant Professor at
the University of New Mexico,
Albuquerque. Received the Rome
Prize in Painting, 1970-72 and
the Alfred Morang Award, 1975.
Lives in Albuquerque.

Solo Exhibitions: Foothills Art
Center, Golden, CO; University
of Colorado, Boulder; University
of Montana, Missoula; Hill's
Gallery, Santa Fe.

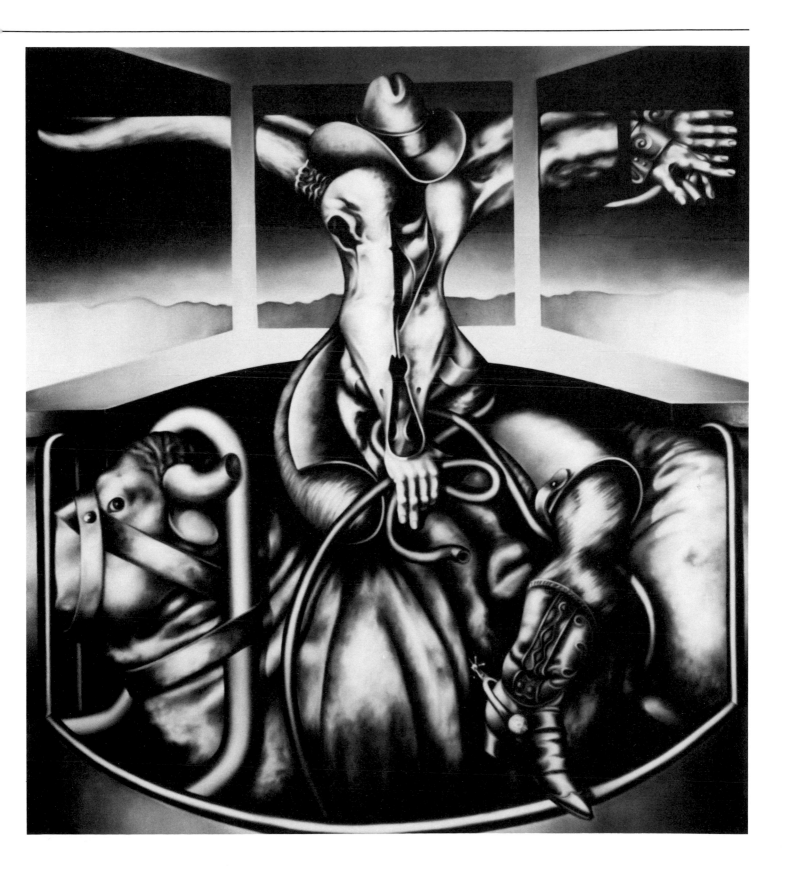

INDEX